# A NATURAL HISTORY OF
# INSECTS
## IN 100 LIMERICKS

The bugman who lived in East Dulwich,
Wrote limericks of such complete rubbich.
He knew he was cursed,
Turning bad rhyme to worst,
So his struggle was poetic justich.

Calvin took pen to paper,
To illustrate books for his pater.
He used, as he will,
His effort and skill,
To put all success to this caper.

# A NATURAL HISTORY OF
# INSECTS
# IN 100 LIMERICKS

RICHARD A. JONES
CALVIN URE-JONES

Pelagic Publishing

Published by Pelagic Publishing
PO Box 874
Exeter
EX3 9BR
UK

www.pelagicpublishing.com

*A Natural History of Insects in 100 Limericks*

ISBN 978-1-78427-250-0 *Paperback*
ISBN 978-1-78427-251-7 *ePub*
ISBN 978-1-78427-252-4 *PDF*

A CIP record for this book is available from the British Library

Typeset by Hewer Text UK Ltd, Edinburgh

# Contents

# Preface

Insects need all the help they can get in the world. They are over-looked because they are so small, ignored because they are deemed trivial, dismissed because they are usually seen as nuisance pests. But in reality they control the world.

Their numbers and diversity are mind-numbing. Back-of-the-envelope calculations give tabloid headline statistics beyond belief – but these are probably all underestimates. There may be 3 million different species of insect out there, mostly in the unexplored rainforests of the tropics. There may be 80 million. Even the experts cannot agree to within an order of magnitude. Their vast numbers and unimaginable variety make them the perfect organisms to study if we want to understand our Earth.

Insects dominate the centre ground of all terrestrial and most aquatic ecosystems. They can tell us the conservation value of ancient woodland and chalk downland. They can show the water purity or pollution level of ponds, streams and rivers. They can help monitor air quality. They can demonstrate the effects of climate change. They offer us a window of unrivalled clarity to look at how the world works. They are warning lights to alert us to the damage that humans are doing to the Earth, and what we can do to try and save it.

Recent insectageddon headlines are starting to make people sit up and take insects more seriously. Insects are vanishing and declining everywhere. But it's not that the warning lights are all going on – it's that all the warning lights are being destroyed. So what better way to promote an interest in these fascinating animals than by poetizing them?

One of my earliest poetry memories is of me, aged about nine, reading out a rhyme I had written about alley cats. Sadly (thankfully) no record of this work now exists. I am a poor poet. It's as much as

I can do to rustle up a bit of doggerel – bad doggerel at that. But I like limericks, and their forgiving frivolity suits me well.

> My fave rhyme is Limerick brevity,
> Five lines and a twist in th'extremity.
> The words quite a mash-up,
> Dr Seuss/Ogden Nash-up –
> Gravitas with a smidgeon of levity.

So – combining the compact nugget form of the limerick with some quite frankly dubious rhyming clashes, I offer this century of discordant nonsense in the hope that entomological outreach will at least benefit from their shock value.

I asked my son Calvin to do the illustrations for this book after he came up with some spectacular one-line drawings in his drop-in after-school art club. I love their simplicity of form, which, along with their linear discipline, echoes the concise rigour of the limerick. He was thirteen years old at the time; he's done spectacularly and I think they're brilliant. Thank you Calvin.

Richard Jones
East Dulwich, January 2021

# Wasp

A wasp with no sting in his tail
Was considered by all a bit frail.
Unlike a sister,
Innocuous mister –
The fate of a *Vespula* male.

It is an indisputable fact of nature that a wasp's sting is a modified part of the egg-laying apparatus – hence only female wasps can sting. Having said this, almost all wasps are females. Amongst the social wasps, *Vespula* and *Dolichovespula* species, the queen that creates the nest from scratch is a female and all of her sometimes many thousands of workers are females too. Males, which have slightly longer antennae and rather blunter abdomens are only reared in small numbers in the nest in late summer, with the next generation of queens. They can be picked up with impunity, to the wonder and awe of the uninitiated – but be careful to ascertain the sex of your wasp correctly first.

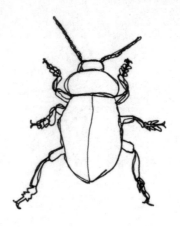

# Bloody-nosed beetle

A bloody-nosed beetle was heard
To let slip an offensive word.
A punch in the craw
Had brought forth red gore,
Making him look quite absurd.

The bloody-nosed beetle, *Timarcha tenebricosa*, is a great lumbering beast of an insect that waddles with a slow clockwork gait. It is too heavy to fly, and indeed it lacks wings. Its defence against predators is to taste foul. Rather than wait to be crushed in beak or maw, though, it reflex-bleeds – exuding a droplet of bright red haemolymph from its mouth if it is disturbed. The copious red liquid is distasteful to birds and animals, and if the beetle is picked up in the hand gives the impression of a bloodied nose.

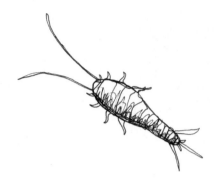

# Silverfish

A silverfish hid in the day,
But skittered at night when at play.
The food that it found
On the floor all around
Was the reason it lived in this way.

Silverfish, *Lepisma saccharina*, are small, delicate, wingless insects, so named for their slender shape, fast movements and covering of silvery scales which slip off easily if they are handled or attacked. They are nocturnal scavengers of spilled food in larders, or in our case under the cooker. Silverfish are considered some of the most primitive of insects; they are wingless and continue to moult their skins throughout their adult lives.

# Fly

A fly shut his wing in the door.
It flipping well hurt, so he swore.
When he tried to flap it
He cried out 'Oh crap, it
Won't bloody well work any more.'

Flies are the supreme aeronauts – otherwise why would they be named after the very act of flying. They are nearly unique amongst flying insects in having only two wings, rather than the more usual four, hence the scientific name of the fly order – Diptera. Amongst the most agile and skilful are the hoverflies (see page 27) and bee flies (see page 58), which can hover stock-still in the air like hummingbirds.

# Earwig

An earwig was more than annoyed
At the fear which his pincers enjoyed.
The weapons he sported
'Are bluff,' he retorted,
'They can't give a nip, take my woid.'

The common earwig, *Forficula auricularia*, is immediately recognizable by the forceps at the tail-end of its body. There is much debate about their purpose. They cannot offer more than a vague nip, but may be useful in startling any would-be predator. There is some sexual dimorphism, since male forceps are strongly curved (as here) whilst those of the female are nearly straight. In some related species they may help in refolding the membranous flight wings, concertina-style, under the short wing-cases, when landing from flight.

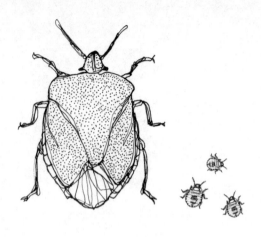

## Shield bug

A shield bug as green as a leaf
Found a frond and laid eggs beneaf.
The nymphs came out waddling,
Bright colours a-modelling.
Protection 'gainst predator teef.

The green shield bug, *Palomena prasina*, is (as with others in the genus) camouflaged an emerald green, and so is well hidden on green leaves. But the small, round, domed nymphs are dark and often strikingly coloured, sometimes spotted. They are presumed to get some protection against predators by mimicking ladybirds, whose bright colours are a warning that they taste foul.

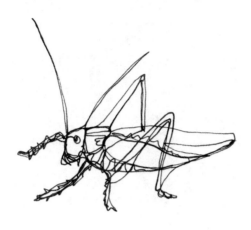

# Bush-cricket

A bush-cricket leapt from its perch,
Pursuing his love in a search.
But one damaged wing
Prevented his sing,
Leaving him quite in the lurch.

Bush-crickets, family Tettigoniidae, sing their courtship songs by rubbing their front wings together. On the underside of the left wing a row of tiny pegs forms a comb which scrapes over a raised ridge (often called the plectrum) on the upper surface of the right wing. Two large areas of the wings, called harp (because of its shape) and mirror (because of its smoothness) act like amplifiers – creating the shrill, loud, penetrating, and sometimes haunting calls.

# Velvet ant

The velvet ant looked soft and cosy,
But under her black, white and rosy,
There lurked a pain-driller
They called a cow-killer.
So certainly not soft and dozy.

The Mutillidae is a family of wasp-like insects, but since the females are wingless and are covered in dense black hair patterned with patches of red and white they have acquired the deceptively friendly English name 'velvet ants'. But they are renowned for having stings powerful enough, supposedly, to kill a cow. Well, not quite, but still very painful. After mating, female velvet ants seek out the burrows of solitary bees and wasps, sneak in and lay their eggs on the host brood. The hatching velvet ant grub then devours its host victim.

# Greenbottle

The greenbottle shone in the sun,
With a sheen that was second to none.
But revolting behaviours
Had done it no favours –
A morning spent feeding on dung.

Greenbottles, *Lucilia* and other species, have spectacularly brilliant metallic green bodies. They belong to the blowfly family, Calliphoridae, which lay their eggs in carrion, animal dung, putrescent fungi or other decaying organic matter. They readily visit dung to feed, sucking up the liquid through a sponge-tipped proboscis. However, they do not come indoors, are not attracted to human food and are not implicated in the spread of bacterial diseases.

# Small copper

The small copper's all of a flit,
Skipping from that flower to thit.
The black and the orange,
A beautiful lozenge,
A creature of marvellous wit.

The small copper, *Lycaena phlaeas*, is a small but brightly coloured butterfly which jinks and skips in rapid energetic flight between flowers or leaf perches. Its caterpillars feed on dock and sorrel leaves, usually in rough grassy places like chalk downs, verges, meadows and gardens. It is renowned for its hyperactive behaviour; even sitting feeding on nectar at a flower it constantly readjusts its position, rotating first one way, then the other, and flitting off in a trice.

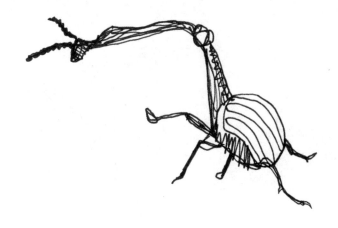

## Giraffe-necked weevil

A weevil had such a long head
That he couldn't fit into his bed.
His neck was so cold,
In his duvet he rolled,
So he slept on the sofa instead.

Males of the giraffe-necked weevil of Madagascar, *Trachelophorus giraffa*, engage in bizarre head-nodding contests. These are ritualized fights and eventually one male will retire, acknowledging the superior head-bobbing ability of his challenger. Females apparently choose the best nodders with which to mate. Females have slightly shorter heads and necks, but still need long necks to cut, manipulate and roll leaf-tubes, in which they lay their eggs.

# Peppered moth chrysalis

A chrysalis started to purge,
And the creature within it emerged.
Instead of a beauty,
The moth was all soooty –
Full black in the background to merge.

During the Industrial Revolution in Britain, heavily soot-stained tree trunks made the pale lichen-patterned form of the peppered moth, *Biston betularia*, stand out against the black. Heavy predation by birds is thought to have reduced the numbers of this original colour form. But the previously rare all-black melanic form of the moth was well hidden and soon became the dominant morph of this species. With the Clean Air Act of 1956, smoke in cities reduced, tree trunks were recolonized by lichens and the typical pale peppered moth once again became the more common form. This is often quoted as a good example of evolution in action. Similar evolution occurred throughout industrialized Europe and North America, following similar soot pollution then clean-air legislation.

# Whirligig

A whirligig beetle felt ill,
With a tension headache to kill.
The cure for her trouble,
On rough neuston bubble,
Was taking a seasickness pill.

Zooming around at top speed on the water, whirligig beetles, family Gyrinidae, form part of the surface-tension community called the neuston. They swim in mad dizzy predator-avoiding gyrations, creating deceptive sun-glinting ripples as they go. They swim using their four short middle and back legs and navigate by feeling vibrations in the surface-tension bubble-like membrane where the water meets the air. The eyes of the beetle are divided and each has split into two, apparently giving the insect four eyes. One pair on top of the head look up into the air above. One pair below look down into the water.

# Trilobite beetle

The trilobite beetles bizarre
Kept all of us guessing – what are
These very strange beasts
That live in the East?
It's adult but looks like larvaa.

Females of the Indian and South-East Asian beetle genus *Platerodrilus* (and others) in the family Lycidae are wingless, and not at all beetle-like. They are neotenic, meaning they retain immature larval features when they become adult. Their strange flattened multi-segmented forms look like the famous fossil trilobites – hence their name. The males, on the other hand, are of typical beetle form, small and elegant, brown or reddish. Little is known of their life histories, but they probably develop as larvae in rotten wood on the forest floor.

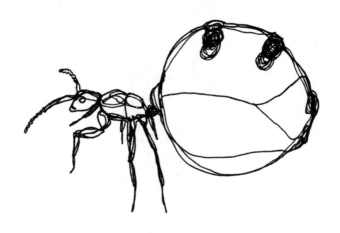

# Honeypot ant

The honeypot sat very still,
Replete, having eaten her fill.
Her belly expanded,
Sugar-coated and candied,
A very confectionary pill.

A few species of ants (e.g. *Myrmecocystus* in the western USA and Mexico, *Camponotus inflatus* in Australia, and *Plagiolepis trimeni* in South Africa), that occur in very arid desert zones use some of their workers to store food. Foragers bring back nectar or honey-dew (aphid excrement) and regurgitate it to the honeypot 'repletes', which become hugely bloated and unable to move. These can then regurgitate some of their stores back to their nest-mates during periods of drought or food shortage. Local people dig up the nests and eat the sweet-tasting honeypots.

## Cochineal

When cochineal insect was pressed,
Her body became rather stressed.
Carmine was the dye for
A colour to die for.
She ended up bloodied and messed.

The cochineal, *Dactylopius coccus*, is a soft-bodied scale insect (related to aphids and mealybugs) in the bug suborder Sternorrhyncha. Originally from Central America, where it feeds on prickly pear cactus, it was deliberately introduced to Africa and Australia, along with its cactus food-plant, because of its high commercial value in producing the red dye carmine. Crushed cochineal was used for dying fabrics (red military uniforms especially), in cosmetics and as a food colour. Artificial chemical dyes made its production less viable in the twentieth century, but it is still farmed and marketed as 'natural red' food colour – although it is not suitable for vegetarians.

# Goliath beetle

Goliath had just enough muscle
For leaf-litter bulldozer rustle.
But public relations
and size connotations
Made everyone kick up a fusstle.

Widely touted as some of the largest insects in the world, African goliath beetles, *Goliathus regius* and *Goliathus goliatus*, can reach 110 millimetres long and about 50 grams in weight. Their life histories are poorly known, but they belong to a group of beetles (chafers) where the grubs usually feed in leaf-litter, fungoid wood or rotting fruit. During the nineteenth century their large size, bright colours and striking forms made them highly collectable, and specimens were regularly sold at natural history auctions for high prices.

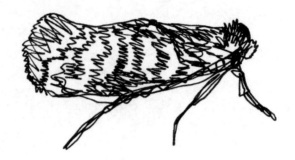

# Clothes moth

The clothes moth has bad reputation
For chewing the shirts of the nation.
But this isn't fair.
The moth wasn't there,
And doesn't possess the dentation.

Several species of clothes moth, *Tinea* and *Tineola* species, are blamed for chewing holes in favourite jumpers. It is not actually the adult moths that do the chewing, as they have only sucking mouthparts – it is their caterpillars. They probably evolved to find nutrition in keratin, a complex protein polymer, in the shed hairs and feathers (and the occasional dead nestling) in animal and bird nests where related wild moth species still occur. Consequently these household pests can only digest animal fibres – so they will infest silk or wool garments but leave cotton and synthetic fabrics alone.

# St Mark's fly

On April the twenty-fifth day,
St Mark left this world for his Dei.
And a heavy black fly,
*Bibio marci*,
Decided to wing on its way.

St Mark's flies, *Bibio marci*, are dark, thick-set, slow-flying flies that bob about rather heavily on the air with legs drooping down and wings heavily shaded with dark clouded markings. They appear, often in abundant clouds, on warm sunny days in spring, usually near St Mark's Day – 25 April, the day on which the eponymous Mark died in 68 CE. The larvae are soil-dwellers, feeding on plant roots.

# Water scorpion

The flat water scorpion had claws
That broke all the natural laws.
He used them for walking,
And also for forking
His food to his undersized jaws.

The water bug *Nepa cinerea* is a 'scorpion' only on the basis that it has a long threatening tail spike, but this is a breathing siphon and not a weapon. The insect is strongly dorsoventrally flattened and looks remarkably like a dead leaf if hooked out in the pond-dipping net. It moves slowly through the water weed, using its strongly angled front legs to capture prey much in the same manner as a praying mantis.

# Cellar beetle

In the cellar a *Blaps mucronata*
Was looking for breadcrumbs to barter.
For years she had gnawed,
Until she was bored,
Eking pudding, and main course, and starter.

The cellar beetle, *Blaps mucronata*, lives in cellars, under floor-boards, in old houses, stables, barns and out-buildings, eking out a living scavenging spilled food scraps. It is long-lived, surviving several years, but is secretive, slow-moving, flightless and rather lumbering in its gait. It should probably be pronounced to rhyme with 'later' rather than 'barter', but I have always had a rather pompous 1950s received BBC English pronunciation in my mind.

# Potter wasp

The potter wasp spoke with a stutter,
Because all the mud made her mutter.
She chewed it in balls
To make her cell walls,
When building her offspring mud hutter.

The potter wasp, *Eumenes coarctatus*, creates delicately sculpted mud flasks for her grubs. Each round pot, the size of a marble, is stocked with small moth caterpillars which she has paralysed, and each receives a single egg before it is sealed up. The larva then feeds on the stored prey. Mud is collected in the wasp's mouth from damp puddles and pond edges, and the small balls of clay are carried back to the nest in her jaws. It takes about 25 mud balls, and 2–3 hours, to make one pot.

# Lacewing

The lacewing lays eggs up on hairs,
Where they sit in the breezes and airs.
Here, predator searches
Confounded by perches
Are the outcome of maternal cares.

Many lacewings, order Neuroptera, lay eggs on stalks. A hair, made of a type of silk, is extruded from the female abdomen as a liquid but solidifies immediately on contact with air, and through the act of stretching it upwards. At the top of each hair a single egg is laid, glued in place. This is partly an anti-predator device, but also discourages cannibalism – since the larvae are ferocious and would certainly eat each other. After hatching they go off, individually, to eat aphids and other small invertebrates.

# Wood ant

A wood ant was called to defend
The nest heap she made with her friends.
She squirted out acid
Until she was flaccid,
By curling her tail round the bend.

Wood ants, *Formica rufa* (and other species), make huge heap nests of pine needles, stalks and twigs, often 1 metre high and 2 metres across. With something in the order of 300,000 occupants, these represent a significant protein source for predators, but the ants will vigorously defend their colony. Each stands back, tucks its abdomen underneath its body and squirts out a jet of formic acid. The jets, reaching 20–30 centimetres, can be directed with some considerable accuracy. Formic acid smells a bit like vinegar and stings if it gets into the eyes, mouth or nose of mammal or bird.

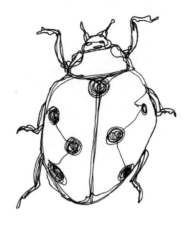

# Seven-spot ladybird

The seven-spot started to tire,
But forgot to cover the fire.
She dozed in a daze,
The house was ablaze.
Lucky no kids in the pyre.

The seven-spot ladybird, *Coccinella septempunctata*, is the default ladybird of folklore and popular culture. The lady after which it is named is the Virgin Mary, who was often depicted wearing a bright red cloak in medieval religious paintings. Ladybirds fly well, and cupped in the hand they are the root of many sayings and rhymes connected with when they fly off – whether this be future lover or betrothed, a wish coming true, or a warning of danger from fire. All good fun, but nonsense, obviously.

# Death's-head hawk-moth

The death's-head pushed into the nest,
Not caring if bees were at rest.
He blew out a whistle,
A sharp peeping hisstle,
Then sucked up the honey they'd blessed.

The death's-head hawk-moth, *Acherontia atropos*, is a sinister-looking beast. However, the dastardly skull outline on the top of the thorax is simply a random pattern, part of the disruptive coloration on its body and wings to help the moth hide, camou-flaged, when at rest. The death's-head hawk's other claim to fame is that it can whistle. It does this by blowing air through its short stout tubular tongue. Although it may take some sustenance from nectar, it is also renowned for using this short proboscis for sucking up honey, and is relatively well known for invading beehives.

# Hoverfly

The hoverfly said 'Look at me,
I'm terribly scary, you see.
I'm all black and yellow,
A dangerous fellow,
I could be a wasp or a bee.'

Hoverflies (family Syrphidae) are spectacular and distinctive insects and many of them are patterned black-and-yellow, -white, -red or -orange. The numerous colour schemes make various species superb mimics of wasps, hornets, honeybees, bumble bees or other stinging insects. The idea that harmless mimic insects get protection by resembling dangerous venomous or poisonous models was best formulated by explorer and naturalist Henry Walter Bates (1825–1892) and is now termed Batesian mimicry.

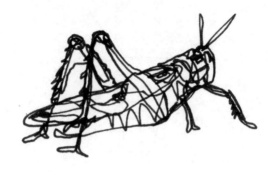

# Locust

The locusts were out for a feed,
But scoured the land clear in their greed.
The fliers and hoppers,
From small'uns to whoppers,
Left famine starvation indeed.

Locusts are large grasshoppers and usually live quiet thinly spread lives feeding on grasses and wild plants. But occasionally they have population explosions. As the wingless locust nymphs (called 'hoppers') feed and grow they sometimes bump into each other. If this reaches a particular threshold rate then instead of becoming a secretive adult, they grow slightly longer wings and larger more powerful thorax so they can fly better – and suddenly they have become gregarious hordes filling the sky in great clouds, blocking out the sun, eating everything they come across and leaving devastation on a biblical scale behind them.

# Oil beetle

The oil beetle tried to take flight,
But her girth gave her bit of a fright.
No matter her straining,
No height was she gaining,
So winglessness now was her plight.

Oil beetles (family Meloidae) have a bizarre life history. They lay tens of thousands of eggs that hatch into tiny active larvae (called triungulins) which climb up plant stems to wait on flowers. Here they grab hold of passing insects. A very few will latch onto the right species of solitary bee and will be taken back to the small nest tunnel in the soil which the bee is stocking with pollen and nectar cake for her grubs. The triungulins move in and devour host larvae and food stores before emerging next year as adult beetles. Female oil beetles are so large, really just bloated bags of eggs, that they have evolved winglessness, relying on their triungulin larvae to disperse to new sites.

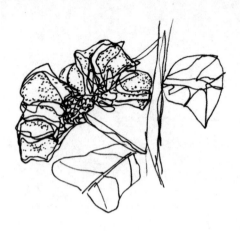

# Knopper

The acorn was growing a wart
Of the sticky and wrinkly sort.
The usurping knopper,
Becoming a whopper,
Reducing the oak seed to nought.

A tiny gall wasp, *Andricus quercuscalicis*, lays its eggs in a develop-ing acorn along with a 'sting' of chemicals that usurp the normal growth process. Instead of an acorn, the tree creates a gnarled sticky canker (the knopper gall) for the grub to feed within. When the species first appeared in the UK in the late 1950s it was feared that it might affect oak production because it destroyed the trees' seeds. Sometimes knopper loads are heavy, with seemingly 100% acorn destruction, but in other years the trees produce plenty of acorns. Acorn loss must be substantial, but is not complete.

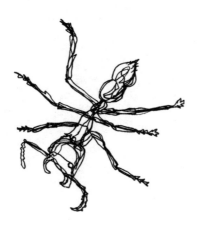

# Bullet ant

The bullet ant feared no attacker,
Because with her sting she fought backer.
The venom she dealt
Was the worst to be felt –
A true hypodermic fire-cracker.

The bullet ant of Central and South America, *Paraponera clavata*, is reputed to have the most painful venom of any insect and is so named because its sting feels like a bullet wound. US hymenopterist Justin Schmidt has published a list of bee, wasp and ant stings, scoring them on a pain scale from 1 to 4. The bullet ant is full 4-point agony, which he describes as 'pure intense, brilliant pain. Like fire-walking over flaming charcoal with a 3-inch rusty nail in your heel.' Ouch.

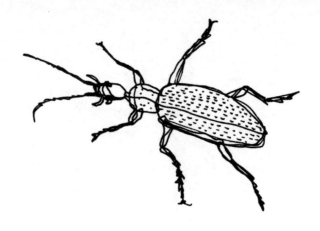

# Bombardier beetle

By mixing peroxide/quinone
In its bottom-end crucible zone,
The exit of gases,
And explosive masses,
Produces a musical tone.

The bombardier beetle, *Brachinus crepitans*, gets its name from the explosive artillery 'pop' (actually a musical 'toot') it makes by combining noxious hydroquinones and hydrogen peroxide in a reinforced crucible chamber in its abdomen. The chemical reaction is strongly exothermic, producing enough heat to vaporize a significant portion of the reactants, which are discharged at up to 100 °C and under pressure. A directional nozzle squirts the boiling hot chemicals and gases out into the face of an attacker – usually an ant.

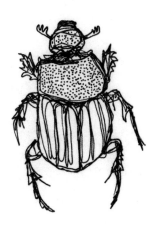

# Dung beetle

A dung beetle dug in the soil,
A life of interminable toil.
But malodorous bait
Would bring him a mate –
The bolus would get him the goirl.

Dung beetles, family Scarabaeidae, are responsible for recycling animal dung, especially mammalian droppings, often by digging holes in the soil and burying portions of it. They lay their eggs in the balls, sausages or boluses of partly buried dung and the grubs feed on this during the weeks or months it takes them to grow big enough to metamorphose into adults. Dung beetles are often smooth and glossy with large powerful broad legs to dig into the ground. Many of the males are armed with head or thoracic spines for push/shove jousts in the underground tunnels.

# Cleg

A cleg bit a horse on the neck
But took something more than a peck.
She drew out the blood,
Thought 'That tasted good' –
But the horse said 'Hey, what the heck?'

Female clegs, *Haematopota pluvialis*, suck the blood of farm animals (horses and cows), and also humans. They need the protein boost to mature a batch of eggs which are laid in the soil. Their larvae are predators after small invertebrates in the leaf-litter and root-thatch. The fly's mouthparts are short and stout and can deliver a painful bite – which is usually the first you know because they are very quiet fliers, and can land undetected. Other species of horsefly (family Tabanidae) also regularly bite people, and although some particularly target neck or legs, *Haematopota* usually goes for the wrist.

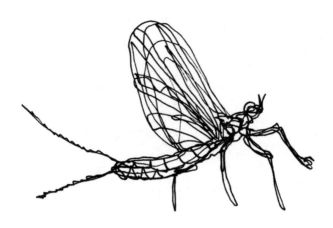

# Mayfly

The mayfly was tired and indignant
That everyone else was so ign'rant.
Though flying one day
Was a span short and gay,
It was nymph-time that made living vibrant.

Mayflies are often depicted as living brief, ephemeral lives; they even belong to the insect order Ephemeroptera. But although the adults may be short-lived, surviving for just a few days or hours, they spend most of their time as aquatic larvae. These nymphs feed on decaying plant matter (a few are predatory). It normally takes them 1–2 years to grow large enough to be able to metamorphose into an adult. The adults emerge, often synchronously, and large clouds of mating mayflies fill the skies.

# Field cricket

A cricket that lived in a hole
Sang a courtship to fit with his role.
But he was alone,
With musical tone,
And none to impress but a mole.

Crickets are well known the world over for their delicate, often haunting courtship songs. In Britain the field cricket, *Gryllus campestris*, is extremely rare, and confined to just a handful of sites in southern England. Captive breeding programmes have been established to increase numbers for release back into the wild, but it remains an elusive insect. Most records of cricket song (often in towns and cities) are from commercially bred house crickets, *Acheta domesticus*, kept as food for pet lizards and tarantulas, which have escaped and which start singing in the early evening on warm summer days.

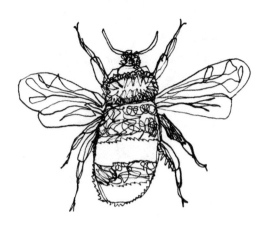

# Bumble bee

The bumble bee got rather hot,
Because she had foraged a lot.
In cold it was good
Having fur coat and hood,
But in sunshine it really was not.

The fluffy bodies of bumble bees, *Bombus* species, enable them to forage in the cold, often before other insects are able to fly. They appear earlier in the year, earlier in the day, further north and further up mountains than other bees. However, they are sometimes in danger of overheating on very hot days. To try and counter this a flying bumble can regulate the flow of haemolymph past a hairless patch on the underside of the abdomen, the heat window, to dissipate some of the excess warmth.

# Dung fly

The dung fly took up his position,
To wait for a mate on a mission.
As soon as she landed,
He pounced, and then handed
A mate-guarding grip proposition.

The golden fluffy males of the common dung fly, *Scathophaga stercoraria*, take up position on a cow pat moments after it is deposited. Here they jostle for position, guarding a territory a few centimetres in diameter, waiting for females. The grey females are pounced upon immediately they appear. After mating, a female lays eggs, but the male continues to ride on her back, guarding her against another male trying to mate. If a female is left unattended another male can mate, scraping out the sperm from previous males before depositing his own. Mate-guarding helps ensures paternity.

# Click beetle

By snapping a jack-knife contortion
The click beetle exercised torsion.
The bend in her middle
Made jumping a diddle,
And clicked out an audible caution.

By tensing a peg on the underside of its thorax against the edge of a pit on the segment behind, and then suddenly releasing it, click beetles, family Elateridae, can suddenly jack-knife their bodies. This alters the centre of gravity so suddenly that the whole insect is jerked 5–10 centimetres into the air. This frequently hops them out of danger, off the log across which they were walking, or down from a flower on which they were feeding. If a click beetle is picked up in the beak of a bird, or between human forefinger and thumb, the clicking is quite powerful, and often startling enough to dislodge the insect, which quickly scurries off to safety.

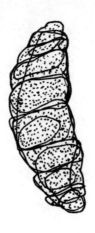

# Warble

The botfly had laid her egg hard
'Gainst the skin of a cow in the yard.
The maggot then burrowed
Through flesh thick and furrowed.
The farmer thought it a bastard.

The cow botfly, *Hypoderma bovis* (also called ox warble), lays eggs on the flanks of a cow. The hatching maggot chews its way under the skin; the wound becomes filled with blood and pus on which the grub feeds. As it grows it burrows through the skin until, nearly fully grown, it reaches the back of the cow, causing large nodular boils called warbles. Eventually the larva exits its gruesome home and drops to the ground, where it pupates. The cows are driven mad by the attentions of the fly, and gad up and down the field to escape – hence its folk name gadfly. The warbles cause serious damage to the cows' health and ruin the leather from the hides. The fly is now very rare in Britain because of an eradication programme using insecticide drenches.

# Hummingbird hawk-moth

The hummingbird hawk knew to fly
At winging top speed through the sky.
She knew how to hover
Without any bovver –
Flew forwards and back, low and high.

The hummingbird hawk-moth, *Macroglossum stellatarum*, flaps its wings so fast they are an invisible blur as it visits flowers. It does not land on the blooms, but hovers perfectly still in mid-air, using its extremely long tongue (longer than its body) to sip nectar. It is very active, flitting about with mesmerizing speed. In Britain it is mostly a migrant, individuals arriving from continental Europe most years, occasionally laying eggs (on its food-plant bedstraws) to form short-lived colonies during the summer and into autumn. There are reports that it can roost during cold snaps, but it seems unlikely that it regularly survives winter here.

# Ground beetle

A ground beetle shiny and black
Had finally mastered the knack
Of wedge-pushing deep
In a warm compost heap,
By the smooth streamlined shape of its back.

Ground beetles, family Carabidae, are named for their habit of running fast across the ground. They also hide under stones and logs, or push through the grass root-thatch. This they can do by the structure of their back legs, which are modified for pushing up and down as well as running fast. The sleek, smooth, wedge-shape of their bodies allows them to squeeze into tight spaces. They then flex special muscles in their back legs which raise and lower their bodies, effectively making their body into a wedge that opens up the crevice so they can move deeper in.

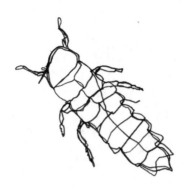

# Glow-worm

A glow-worm from outskirts of Brighton
Had a sore tooth she tried to shed light on.
She used some contortion,
But lacking in caution,
Extinguished herself with one bite on.

The glow-worm, actually a beetle, *Lampyris noctiluca*, is the wingless worm-like female of the species. She has a glowing light-producing organ under the tip of her tail that she uses to signal her whereabouts in the grass so that the winged males can fly down to find her in the night. A complex chemical called luciferin is activated by an enzyme, luciferase, to release a light particle (a photon), but without generating any heat. This 'cold' light is the exact opposite of photosynthesis, where chlorophyll in plants absorbs photons to combine carbon dioxide and water to create sugars.

# Stag beetle larva

A stag beetle maggot had grown
To completion, as far as he'd known.
He stopped and pupated –
But he should have waited.
He's tiny – he's all skin and bone.

Stag beetle, *Lucanus cervus*, larvae feed in rotten fungoid logs and stumps, but the low nutritional quality of their food means they have to spend several years chewing away until they are large enough to pupate and then emerge as an adult. Males in particular need to reach a size that will enable them to grow the huge antler-like jaws during the metamorphosis from larva to adult. Well-fed larvae produce huge beetles, up to 50 millimetres long (not including the jaws), but malnourished grubs can produce miniature males just 20 millimetres long. Small males benefit by having shorter larval lives, so they reduce their ongoing risk of being eaten by predators or succumbing to fungal disease, but they cannot compete with larger males who have waited longer.

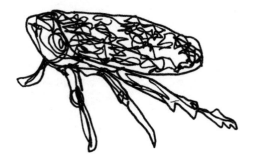

# Spittle bug

A froghopper nymph lived in spit,
But was not at all happy with it.
Although the soft foam
Constituted a home,
He was fed up, and ready to quit.

Cuckoo spit has nothing to do with cuckoos, but is a wet froth made by the pale, soft and vulnerable nymphs of jumping frog-hoppers, *Philaenus spumarius*, also known as spittle bugs. The nymphs suck plant sap and pass copious liquid waste which they froth up by blowing bubbles into it. The white foam hides them from predators and parasitoids, protects them from desiccation in hot weather, and keeps them warm in cool weather. Eventually they pupate inside the foam, which becomes stiff and dry, and finally the winged hopping adult emerges.

# Deathwatch beetle

The deathwatch is lost, so it's said;
On heavy oak beam he has fed.
In the cold, dark void,
He can't hear a woid,
So takes to head-banging instead.

Deathwatch beetles, *Xestobium rufovillosum*, breed in the rotten heartwood of old trees. If they occur indoors it is because their ancestors were brought into the house in the construction timber when it was first built, often centuries before. Many generations later, their tunnels can combine to create fragile honeycomb labyrinths and voids which will eventually weaken the beams to the point of collapse. In the darkness of these voids the beetles find each other, to mate, by banging their heads on the wood, creating the ominous tapping sound of a clock or watch. In the still silence of the night this tapping, coming from the very fabric of the building, seems a portent of doom or death, ticking away the hours.

# Minotaur beetle

The minotaur beetles dug down,
Under faeces all fragrant and brown.
They buried the crots
As food for their tots,
Two tunnellers of great renown.

The minotaur beetle, *Typhaeus typhoeus*, is named for the long bull-like horns that grow from the front edge of the male thorax. These are lacking in the female. Usually a male and female pair work together to dig a deep burrow near animal dung – often cow or horse droppings, but also deer fewmets or rabbit crottels. They are renowned for digging very deep burrows in sandy soil, often reaching over a metre down. A sausage-shaped stash of dung is impregnated with eggs, and the beetles' larvae feed on the stored excrement. Although unobserved, it is likely that horned males joust with each other in the underground tunnel, the victor gaining control of burrow, female and dung store.

# Tiger moth

The tiger moth, patterned and bright,
Has suffered a terrible plight.
No matter the season,
Whatever the reason,
It's always attracted to light.

It is still uncertain why the garden tiger moth, *Arctia caja*, indeed most moth species, are attracted to lights at night, even though this behaviour has been known for millennia. It seems unlikely that they are suffering navigational confusion, because moths are not known to use the moon to orient themselves, or for plotting migratory trajectories. It is possible that the visual sensors in their eyes become overloaded by the bright light, such that nerve signals are no longer transmitted to the brain, giving the impression of a dark area in their field of vision, into which they try to fly to seek shelter. Lights are a real danger to moths, which are likely to end up immolating themselves in the naked candle flame, bashing themselves to exhaustion against the window, or being picked off by bats flying around the street lamps.

# Caddis

The caddis, a master with bricks,
Made a case of snail-shell mix.
The spiral designs,
On classical lines,
Was one of its cleverer tricks.

The aquatic nymphs of caddisflies, order Trichoptera, make tubular cases incorporating sand grains, tiny pieces of gravel, short pieces of plant stem, fragments of dead leaf, or sometimes empty snail shells, all held together by silk strands. Each caddis species specializes in creating a distinctive size and shape of case, using its own choice of particular materials.

# Deer ked

The deer ked is flat in the face –
Indeed, flat all over the place.
The reason is clear
In the fur of the deer,
Through which it can crawl at a pace.

The deer ked, *Lipoptena cervi*, is a small blood-sucking fly that specializes in attacking deer. When it lands on its host it sheds its wings; unencumbered, it then pushes through the fur, down to the skin to feed. It is dorsoventrally flattened, to more easily push through the animal's thick pelt. A female incubates one larva at a time in her abdomen, retaining it in a womb-like organ where it feeds on proteinaceous secretions from its mother. Only when the larva is fully grown is it released; it drops to the ground and pupates immediately. When the winged adult emerges it must fly off to find a new host animal.

# Cinnabar

A cinnabar moth, red and black,
Acquired and practised the knack
Of flying by day,
Not hiding away,
Rejoicing, and not looking back.

The cinnabar moth, *Tyria jacobaeae*, is a day-flying moth, often mistaken for a butterfly. It is brightly coloured, a strong warning to would-be predators that it is distasteful. The equally strongly marked black-and-yellow barred caterpillars feed on ragwort and sequester toxins from the food-plant, which are retained through metamorphosis into adulthood.

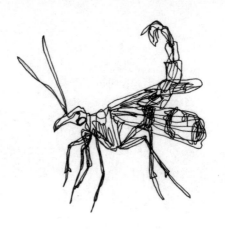

# Scorpionfly

The scorpionfly was a sight,
With a tail of muscular might.
But no actual sting
Would come from that thing.
'Twas a case of scorpio lite.

Scorpionflies, order Mecoptera, family Panorpidae, are named for the male's large swollen sting-like tail, which curves threateningly up from behind the wingtips. But the reddish pointed bulb is merely the bizarre male genitalia; females are unarmed and have dull tubular tails. It is difficult to suggest that the insect gets any protection from scorpion mimicry, because only one sex has the bluff, and these insects occur in many temperate zones where scorpions do not live. Adults and larvae are scavengers, feeding on dead plant material, and sometimes on dead insects, even those found in spider webs.

# Conopid fly

The conopid fly was not glad.
Body image was making her sad.
Big-bottomed, they said,
And thick in the head.
No wonder behaviour was bad.

So-called 'thick-headed' flies (family Conopidae) have broad heads, and bulbous swollen abdomens. Their large eyes allow them to find their host victims – bees, wasps or hornets – which they grapple in mid-air. The large tail tip, especially of the females, contains a ferocious egg-laying blade (likened to a can-opener in some reports) that allows them to cut into the bee or wasp body, prising open the abdominal segments, to lay an egg. The grub then hatches and eats its victim alive, from the inside.

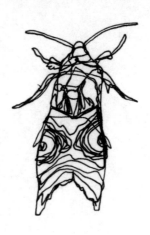

# Angle shades

Sporting art deco design,
On each wing an angular line,
The angle shades rest,
With camouflaged vest,
The leafest-like mimic you'll find.

The striking angular art deco pattern of beige, gold and pale green found in the angle shades moth, *Phlogophora meticulosa*, is very pretty in museum specimens, but in life, resting on the foliage, the moth looks just like a curled and twisted dead leaf. Though the angular patterns are strikingly stark, they still look very naturalistic, because they break up the smooth silhouette of the moth when viewed sideways-on, as it would be by a predator bird, hopping about through the undergrowth.

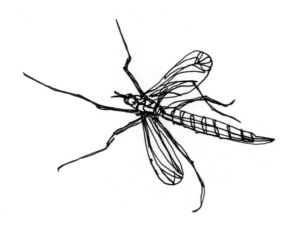

# Cranefly

A cranefly of very long lank
Half flew and half crawled up the bank.
His flight was a wobble,
And it made him hobble,
As into the long grass he sank.

Craneflies, family Tipulidae, are amongst the lankiest of all insects, with legs often many times longer than the body. Sometimes called daddy-longlegs, this can give rise to confusion with other equally long-legged creatures such as harvestmen (arachnid order Opiliones) and the daddy-longlegs spider, *Pholcus phalangioides*. The long legs are fragile and easily broken, which may be a defence against predators since the insect can survive even after losing several limbs. Very characteristically, they half fly, half crawl through the long grass, using their long legs to drag themselves about, whilst their narrow wings barely provide enough lift to support their body weight.

# Mother Shipton

The Shipton moth flies in the sun,
Unlike those that night-fly for fun.
Its pattern is clear –
A mythical seer,
Misquoted, invented, undone.

The wing pattern of the Mother Shipton, *Callistege mi*, apparently shows the craggy face of an old hag, complete with beady eye, toothless mouth, hooked nose and prominent chin. This is supposed to represent the fifteenth/sixteenth-century witch and fortune teller Ursula Southeil, or Southill, or Soothtell, better known as Mother Shipton. A number of prophesies were attributed to her – including the end of the world, horseless carriages and steamships – but these are clear fabrications, published centuries after her death.

# Robber fly

*Asilus* launched after her prey
From a dung pat on each sunny day.
With demeanour ferocious,
And mouthparts atrocious,
Her victim seldom got away.

The hornet robber fly, *Asilus crabroniformis*, is one of Britain's largest flies, and certainly its most ferocious. Typically, it sits on a patch of bare soil or a dry cow pat and launches itself at passing flying insect prey. Using its sharp and powerful mouthparts, it can puncture the shell of dragonflies, large dung beetles, wasps, and other insects often much bigger than itself. Its black and yellow colour scheme, highly reminiscent of a hornet, also makes it perhaps Britain's most sinister-looking fly, though it is completely harmless to humans.

# Bee fly

With body a bundle of fluff,
And a tongue which is just long enough,
It visits a primrose,
And inserts its long nose –
The bee fly's a live powder puff.

The common bee fly, *Bombylius major*, is a true herald of spring, emerging in March and April. It is frequently seen hovering motionless in mid-air drinking nectar from primrose flowers with its long proboscis, its wings an invisible blur. It lays its eggs in the burrows of solitary bees. This it does by ejecting an egg in mid-flight, whilst hovering over the bee's entrance hole. The grub that hatches devours the bee brood and the store of nectar and pollen that the host bee has packed into the burrow.

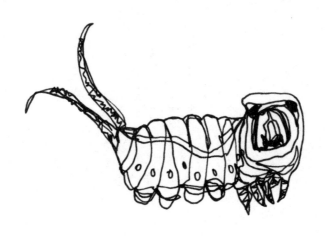

## Puss moth caterpillar

Presenting a square-ended head,
The puss moth cat'pillar looks dread.
But beware of the tail –
It squirts formic ale,
Which smells vineg'ry, so it's said.

The caterpillar of the puss moth, *Cerula vinula*, is a brightly coloured green and brown creature, with square face and strong saddle mark on its back. The markings help break up its silhouette, hiding it from the usual bird predators which hunt by sight in the herbage. If it is attacked it has one last line of defence – it can squirt formic acid (very similar to acetic acid, vinegar) from its tail flagellae.

# Picture-winged fly

The picture-winged fly had a drawing
On each aerofoil of its forewing.
The dots and the dashes,
Made patterns and splashes –
A dipteron definitely not boring.

Picture-wing flies (families Tephritidae, Ulidiidae, etc.) have markings across the wings – bars, dashes, stars, speckles, dots. These are fascinating and beautiful. In some species they seem to play a part in semaphore communication signalling between the sexes during courtship. In others they may give camouflage by breaking up the outline of the wings. In yet others they may create distraction patterns by looking like false heads or the silhouette of an ant.

# Rose chafer

The golden rose chafer aglow,
In the afternoon light long ago,
Made a wonderful sight
As it took off in flight,
Buzzing soft in the air as it go.

The rose chafer beetle, *Cetonia aurata*, is a jewel of beaten metal, a polished golden green. Most beetles can fly – they flip open their elytra (wing-cases), and extend their folded membranous flight wings. *Cetonia* (along with other chafers) has a notch on the side of each elytron so that after its flight wings are extended it can reposition its wing-cases back down over its body.

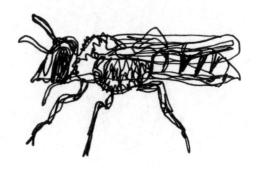

# Bee-wolf

*Philanthus* has such a sweet tooth,
But her bee killing's seen as uncouth.
She buries her prey
For her offspring today,
Who need protein tomorrow forsooth.

The bee-wolf, *Philanthus triangulum*, is a solitary wasp that specializes in attacking honeybees. It takes the paralysed victims back to its burrow and places each bee in a small chamber off the main tunnel. On each it lays an egg. The hatching maggots eat their honeybee hosts before pupating. Up to 30 bees may be buried along a tunnel nearly a metre long dug down into the sandy soil.

# Small tortoiseshell

The tortoiseshell basked in sun light,
But thought that he'd have him a fight.
Whilst looking for dames,
He challenged the planes,
The bumbles, the birds and a kite.

Male small tortoiseshells, *Aglais urticae*, guard a patch of sunlight as their own personal territory. They fly up to investigate other butterflies passing through, to court a female in a whirling aerial dance, or joust with a male, often making wing contact. Their eyes do not offer very good resolution, but are very sensitive to movement – so they will also fly up to investigate other butter-flies, birds, bumble bees, and even the occasional passing aero-plane. To determine if a basking tortoiseshell is a male, simply lob a small stone about a metre above it. A male will challenge the stone; a female will ignore it.

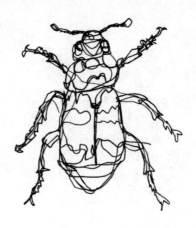

# Burying beetle

The burying beetles were drawn
To a dead pigeon out on the lawn.
They dug in the soil
Till, buried in spoil,
The odorous carrion was gawn.

Also called sexton beetles, after the attendant in charge of ceme-teries and graveyards, burying beetles, *Nicrophorus* species, do just that – they bury. Attracted to the scent of decay, a pair of beetles, male and female, excavate the soil around a small carcass until the hole consumes the corpse and the spoil heap covers it over. They then lay eggs and tend their offspring by chewing up some of the putrefying flesh and feeding it to the larvae.

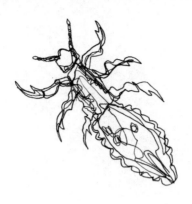

# Head louse

A head louse from Berwick on Tweed
Took a wrong turn when trying to feed.
Instead of the crown,
He went further down,
And was squashed in the groin for his greed.

Head lice, *Pediculus humanus*, feed (by sucking blood) only on human heads, and get transferred from one person to another only by physical contact – brushing heads together during cuddles and huddles. They have large claws precisely adapted to hanging on to human head hair. The crab louse, *Pthirus pubis*, on the other hand, is the louse of coarser body hair, especially the pubic area. Usually spread by intimate sexual relations, *Pthirus* is the most embarrassing of all insects. Sadly, head louse infestations are sometimes stigmatized, but they will feed on all heads – clean or dirty, rich or poor.

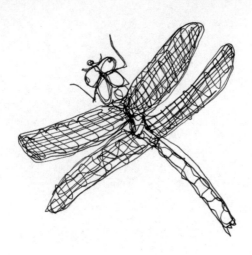

# Dragonfly

A dragonfly once caught a gnat,
And thought 'I like the taste of that.'
He then caught another,
Its sister or brother.
The whole family later, he's fat.

Dragonflies (suborder Anisoptera, within the order Odonata) are supreme aeronauts, catching their flying insect prey in mid-air. They have huge compound eyes, each with nearly 30,000 facets. Each facet has a lens across the top and light-sensitive cells in a column beneath. This gives a pixelated view of the word, but one which is very sensitive to relative movement, allowing the dragonfly to calculate precise interception trajectories in the air.

# Orange-tip

An orange-tip flew through the glade,
And on lady's-smock food-plant she laid.
But one caterpillar
Per plant, or the killer
In each, is the price to be paid.

The orange-tip butterfly, *Anthocharis cardamines*, lays its eggs on lady's-smock, *Cardamine pratensis* (also known as cuckoo-flower), a plant of damp woodland edges, hedgerows and road verges. Only one egg per plant is laid, because the newly hatched caterpillars are voracious and easily take up cannibalism if they discover another egg or a smaller larva nearby.

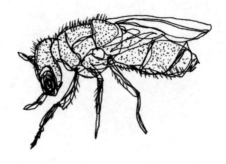

# Ruby-tail wasp

The ruby-tail wasp would not settle,
Her fidgeting was in fine fettle.
Green- and red-tinted,
She flitted and glinted,
All dinted and shining like metal.

Ruby-tail wasps (family Chrysididae) look as if they are made out of beaten metal. Often shining golden, red or green, they are constantly moving, confusing the eye with their bright reflections and glints. They are cleptoparasitoids, sometimes called cuckoo wasps. They lay their eggs in the burrows of other wasps which have dug a tunnel into soil or dead wood and laid up stocks of dead insects for their grubs. The ruby-tail grubs hatch first and devour both the host larvae and the food stores.

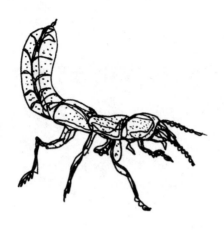

# Devil's coach-horse

When *Ocypus olens* is vexed,
Feels threatened, or frightened, or hexed,
He oozes a drop
Of faeces on top
Of his hind-body – upwards reflexed.

The devil's coach-horse, *Ocypus olens,* is Britain's largest rove beetle and a most gothic-looking insect. Rove beetles have short wing-cases, leaving their flexible hind-body exposed. This gives them sinuous manoeuvrability in tight spaces under logs and stones or through the root-thatch. *Ocypus* has large jaws and can give a slight nip if handled incautiously, but when threatened it curls up its tail, scorpion-like, exuding a droplet of foul-smelling faeces to daub into the face of any would-be predator.

# Magpie moth

The magpie moth, black spotted white,
Tried to take on a bird in a fight.
But avian brain
Thought 'No not again',
And refused to have even one bite.

The magpie moth, *Abraxas grossulariata*, is strongly marked black on white, usually with suffused bars of orange. Its striking colours warn birds that it is distasteful, and they are loath to eat it. Reasoning goes that if a bird eats one brightly coloured but foul-tasting insect it soon learns to avoid all other similar items.

## Purple emperor

The emperor landed in style,
But his regalness still raised a smile.
His tipple of choice
Was not very noice –
Dark liquor from excrement pile.

Renowned for its glorious colours and great rarity, the purple emperor, *Apatura iris*, is also well known for its strange tastes. It does not visit flowers to take nectar. Instead it drinks the liquid from fermenting fruit, oozing fungoid sap from damaged trees, fresh animal dung, carrion, car exhaust pipes, or rancid baits made by butterfly-watchers from ripe cheese or prawn paste.

# Stonefly

A stonefly of tubular form
Made freshwater his habitat norm.
The slightest pollution
Was full persecution,
A poisonous ruinous storm.

Stoneflies, order Plecoptera, have rather flat or cylindrical adults which furl their four wings tight onto the body. This makes it easier for them to push down and hide in the waterside vegetation. They have aquatic nymphs which are particularly sensitive to water pollution. The richest, most diverse stonefly faunas are only found in the purest running freshwater streams, and their populations can be used to monitor water quality.

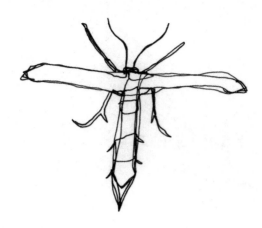

# Plume moth

The plume moth sat stiff on the grass,
Waiting for danger to pass.
Its pale constitution
A clever solution,
Giving it elegant class.

Plume moths, family Pterophoridae, have long narrow feathery wings that they furl into stiff rods held out at right angles to the body when at rest. They are pale or drab dead-leaf colours and look very angular, stiff, almost twig-like. They presumably get some camouflage protection from predators because of it.

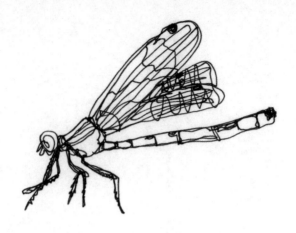

# Damselfly

The damsel that flittered so light
Could still deliver a bite,
To take out a midge
Flying over the ditch,
And give all the mozzies a fright.

Though small, delicate and flimsy, damselflies (suborder Zygoptera, within the order Odonata) are serious predators – but of small, delicate and flimsy prey like gnats, midges and mosquitoes. Unlike their dragonfly relatives, which hunt in the open air around ponds, lakes and rivers, damselflies flutter very close to the water's edge, flying through the tall vegetation. They often snatch at prey resting on the waterside foliage, rather than grabbing flying insects.

# Cockchafer

The cockchafers flew out in May,
At dusk, at the end of the day.
They crashed into lights
On warm sultry nights,
But then they all vanished away.

Cockchafers, *Melolontha melolontha*, fly about in May, and are sometimes called maybugs. They are attracted to lighted windows and street lamps and they crash noisily against the glass. The larvae are grass-root feeders and they used to be considered agricultural pests. Mass emergences could turn the sky grey with their clouds, and in 1574 so many drowned in the River Severn that they clogged the water wheels of the mills. Changes in agriculture mean that cockchafers are nowhere near as common as they once were.

# Comma

The comma has two distinct forms,
Which goes against butterfly norms.
The dark hibernate,
But the pale cannot wait,
And a second brood often performs.

The comma butterfly, *Polygonia c-album*, has two underside colour forms. Some spring caterpillars feed up quickly and emerge as May and June butterflies with paler brown undersides. These mate and lay eggs for a second generation which flies in August. Other spring caterpillars feed slowly and do not develop into adults until late July or August and have dark, sometimes almost black, undersides. Both forms hibernate as adults inside hollow tree trunks, rock crevices, ivy thickets or sheds.

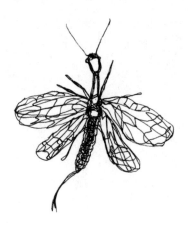

# Snakefly

In woods a primordial insect
Has bulbified head and is long-necked.
The snakefly's a goth
Compared to a moth,
And its egg-laying tube demands respect.

Snakeflies, order Rhaphidioptera, are an archaic group of insects related to lacewings and beetles. They have very beetle-like larvae which hunt small invertebrates in rotten wood or leaf-litter. Not many species are known – about 250 worldwide, only four in the British Isles. Despite their startling appearance and long tail spike, snakeflies are completely harmless to humans.

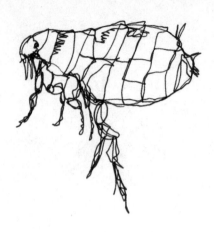

# Cat flea

For cat fleas life's now a beginner,
Because in cocoons they've got thinner.
In carpet they hide
As you come inside –
They've decided to have you for dinner.

Adult cat fleas, *Ctenocephalides felis*, suck the blood of their hosts, but they drop their eggs into the carpet where the tiny worm-like larvae feed on flea droppings (effectively dried blood) and various other bits of detritus. When fully grown they pupate, but remain dormant and do not emerge as adults until they feel the vibrations of the animal returning to the 'nest'. If the house remains empty for a holiday period the fleas all wait in their cocoons, and within minutes a mass synchronous emergence of plague proportions greets the first returning householders to come into the room.

# Speckled wood

The speckled wood had just expanded,
On Scottish hillsides it highlanded.
It favoured wood glades,
With sun-dappled shades
That its grass-feeding larva demanded.

The Speckled Wood, *Pararge aegeria*, is a denizen of woodland
rides, forest edges, hedgerows, verges and other areas of mottled
shade, where its caterpillars eat various common grass species. It
has long been widespread in England and Wales, but in the last
seventy years has started to expand up into Scotland and also
across most of Ireland. The reasons for its change in fortune are
uncertain, but climate change and loss of traditional coppice
woodland management, leading to more shady woods, seem
likely to have helped it.

# Termite

The termites built up to the sky,
To watch at the sun passing by.
They built very nimbly
A vent in the chimbly,
To help take waste gases up high.

Many mound-building termite species in Africa, Australia and South America build large nests, using saliva to cement soil particles together into a hard concrete-like structure. Although historically sometimes called 'white ants', termites are not ants, but actually social cockroaches. They belong to the suborder Isoptera, within the order Blattodea. In some species the nests are oriented north–south, to catch the morning easterly sun on a broad side, but present a narrow profile at noon so as not to overheat. Inside, tunnels and chambers allow warming air to rise and escape through vents, taking waste gases up and away, but drawing in fresh air from the edges. A kind of engineered climate control.

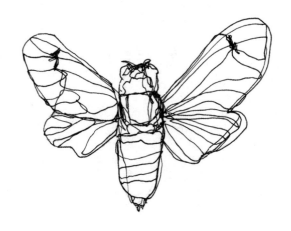

# Cicada

After seventeen years in the ground,
The *Magicicada* looked round.
Dozens and dozens
Of all of his cousins
Were singing – and wow, what a sound.

The North American periodical cicada, *Magicicada septendecim*, spends seventeen years as a root-sucking nymph before hauling itself up a plant stem and emerging as a winged adult. They now start 'singing' to mark territory and woo a mate. They then lay eggs and die. Mass emergence after a gap of seventeen years seems like magic. The seventeen-year life span is thought to throw off parasitoids and predators, which cannot compute this prime-numbered cycle. Different regions have different cycles; the largest (generation X) occurs in north-east USA from New York to Virginia, Illinois and Michigan. Its last appearance was in 2004 and the next are due in 2021 and 2038.

# Bed bug

A bed bug hid under the sheet,
Waiting for legs and for feet
To bite and to suck,
And then with some luck,
To hide up, remaining discreet.

The bed bug, *Cimex lectularius*, is a flattened wingless insect that hides in the cracks of bed-frames, behind loose wallpaper, under carpets, or tucked up in the mattress. It emerges at night to suck blood, then hides again during the day. It was almost eradicated in the days of DDT, but has recently enjoyed a resurgence. Modern sensibilities mean people want to treat it with chemicals, but these are less powerful than the now banned DDT, and they are less willing to take beds apart, take up carpets and explore crannies to squash the bugs, even though this is the best solution to the problem.

# Water skater

The skater took very great care
To stand on the water and stare.
On look out for ripples,
From wave crests and dipples –
Its target prey better beware.

Water skaters, *Gerris* species, spread their weight on long thin legs and long feet so that they are supported on the water surface tension, where they oar themselves around at top speed. They sense the vibrations made by other small creatures in and on the surface, including small insects drowning. They use their front legs to manipulate prey and have long skewer-like mouthparts to attack their victims.

# Praying mantis

A mantis, refusing to pray,
Spent a hungry few hours one day.
He claimed to be pagan,
But took home no bacon,
Then saw the err'r of his way.

Holding its front legs folded as if praying, the European mantis, *Mantis religiosa*, is a sit-and-wait ambush predator. Remaining stock-still until an unsuspecting fly or beetle comes close, it then snatches out with its powerful clawed limbs and grabs its prey out of the air.

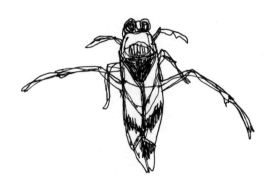

# Water boatman

The boatman took care of its oars,
To make it one of the best rowers.
But swimming below
Did not make it slow,
Even when taking a pause.

Water boatmen, *Notonecta* and *Corixa* species, use their long, fringed back legs as oars to swim through the water. They are remarkably rapid. However, they often rest quietly just below the water surface, hanging by their tail-tip. Here they are replenishing their air supply, which they carry in a thin bubble film around the body. If disturbed by a movement, or a shadow passing over them, they suddenly flick into action, and disappear quickly down into the depths.

# Cockroach

The cockroach ran fast as the wind,
In search of old food that was binned.
But he found it harder
Invading the larder,
Where everything ed'ble was tinned.

Cockroaches, order Blattodea, are renowned, if reviled, domestic pests, scavenging on food spilled in the pantry and under the cooker, or by raiding bins. They run very fast and can hide in small spaces under kitchen appliances or cracks in skirting boards. They are somewhat less common in developed countries, where refrigerators, Tupperware and tinned goods keep them out. Cockroaches are among the fastest recorded running insects, clocking up speeds of 1.5 metres per second.

# Leafcutter bee

With nothing to measure the curve
Of circles and ovals, to serve
The segments of leaf
She cut with her teef,
The leafcutter chewed on the swerve.

Leafcutter bees, *Megachile* species, cut neat semi-circular or oval fragments of leaf to line the small tunnel nests they make in the soil, burrowed into dead wood, or in ready-made holes such as hollow stems, small pipes or wind chimes. The leaf fragments are expertly engineered and constructed into cigar-like tubes ready to take stores of nectar and pollen cake for the grubs.

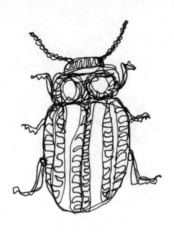

# Rainbow leaf beetle

The rainbow leaf beetle knew well
That his rarity could ring a bell.
There's a small population
In a mountainous nation –
A secret life up in the dell.

The rainbow leaf beetle, *Chrysolina cerealis*, occurs (in the UK) only in a few sites in Snowdonia, north Wales, where its larvae feed on stunted wild thyme plants. It is very beautiful – a metallic golden red dome with several bands of purple edged with green. It is so rare that it is one of the few insects protected by UK law against wilful disturbance or destruction of its habitat.

# Ichneumon

The ichneumon gave a small grin,
As she laid her first egg within.
The hornworm had flinched,
The stinging had pinched,
Then nothing? – the prick of a pin.

Ichneumon wasps, family Ichneumonidae, are a large group of insect parasitoids. They mostly lay their eggs in butterfly and moth caterpillars. Some of the largest target the caterpillars of hawk-moths, called hornworms for the sharp spine at the tail end. The ichneumon grub hatches out and eats the caterpillar alive from the inside. The host caterpillar may live for some days or weeks, oblivious to the larva living inside it, and it will carry on feeding, continuing to provide increasing sustenance to its parasite. It may even pupate, but only the adult ichneumon will ever emerge from the chrysalis.

# Mosquito

Mosquito flew light on the air,
Descending to sample blood, where
O, A, B, AB,
Doesn't matter to she,
Who sucked it with nary a care.

Only female mosquitoes suck blood. They need the extra protein nutrition to mature a batch of eggs – which they lay in ponds, ditches, puddles or other small waterbodies. Although some people claim to be bitten more often (or less often) than the norm, blood type, skin type, complexion, body shape, age and gender seem to have no effect on how the mosquitoes find their victims, or on the voracity of their attack.

## Goat moth caterpillar

The goat moth was sick of the smell
That spoke of a capricorn hell.
Its maggot had chewed
In deep fungoid woood,
With beetles and fly grubs as well.

The reddish-pink and brown fleshy caterpillar of the goat moth, *Cossus cossus*, chews burrows in large old tree trunks, particularly sallows, willows and poplars. The sap, caterpillar droppings and fungal infection seeping into the damaged wood create a strong-smelling ooze supposedly reminiscent of goats – hence the insect's name. Many other dead-wood-feeding fly and beetle grubs then also invade the large rotting odorous wound on the still-living tree.

# Horntail

A horntail of panic proportions
Should have been issued with cautions.
The spike on her tail
Made everyone flail.
She ended up swatted – in portions.

The horntail, *Urocerus gigas*, is a large sawfly, so named after the saw-toothed egg-laying tube it uses to drill into dead tree trunks. The insect is a large black and yellow wasp mimic, but the spike at the end of its tail contains no sting. It is harmless, if a bit menacing. The long-lived larvae develop in dead wood, and are sometimes accidentally brought into buildings when they are contained in apparently sound construction timber. Many months later they emerge from the wooden floorboards or joists, causing panic amongst the unwary householders.

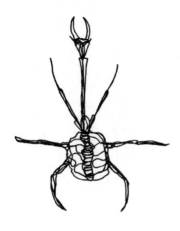

# Antlion

Antlion sat deep in the pit,
Patiently waiting till it
Could skewer the ant
That, ignorant,
Lacked vigilance, caution and wit.

Antlions are the ferocious larvae of rather delicate fluttering adults (family Myrmeleontidae) resembling lacewings. The larva digs a conical pit in dry, crumbly sandy soil and waits at the bottom for ants to fall in. If the ant struggles to escape the antlion flicks up sand grains to cause an avalanche, bringing its doomed victim down into its large jaws. This long-necked 'violin larva' is an eastern Mediterranean species called *Dielocroce hebraea*. Most antlion larvae are short and squat.

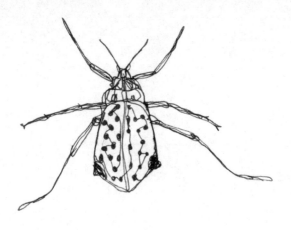

# Aphid

An aphid was sucking plant liquor,
When her neighbours all started to bicker.
They waggled their tails,
Back legs whirred like flails,
To fight off attacker by kicker.

Aphids (also called greenfly or blackfly, depending on their colour) are sap-sucking plant lice. With a long thin proboscis buried deep into a plant leaf, they can't always just get up and walk away if a predator such as a ladybird or lacewing larva comes near. By releasing an alarm pheromone, an aphid being attacked can orchestrate its neighbours to start bucking about and kicking with their back legs to try and jointly ward off the enemy.

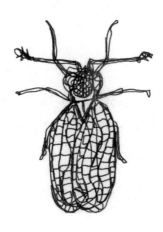

# Lacebug

A lacebug complained it was silly
That she should be dressed up so frilly.
Her delicate form,
Reticulate norm,
Reduced her to tears, it did really.

Lacebugs (family Tingidae), are very small (most less than 5 milli-metres), but beautifully sculpted when viewed under the micro-scope. Their bodies appear to be decorated with wrinkled and netted patterns resembling lace. The exact significance of this is not clear, but the bugs may gain some protection from predators because they are not shiny like most insects, and the corrugation may help them blend in with the reticulate texture of the leaves on which they are feeding.

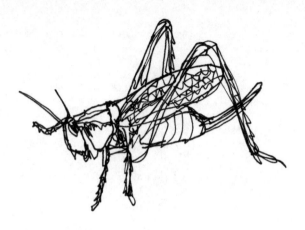

# Wart-biter

The wart-biter service was real
Good help as it chewed at its meal.
It wasn't too tasty,
In fact it was naisty –
Verruca grown hard on the heel.

The wart-biter, *Decticus verrucivorus*, is a large bush-cricket. Like most bush-crickets it is omnivorous, catching and eating small insects as well as nibbling plants. Its large powerful jaws can deliver a painful nip, easily breaking the skin if handled incautiously. There is, reputedly, a folk tradition of using these insects to chew off warts, verrucas and other areas of tough thick skin from the hands and feet.

# Stick insect

The stick insect sought out a mate,
So he could go on a date.
But she didn't need,
His input of seed –
No need to imperegnate.

In many species of stick insect, order Phasmatodea, males are very rare or unknown. Females are able to lay fertile eggs without mating, through a process called parthenogenesis. In most organisms eggs are produced by splitting the number of chromosomes (carrying the genetic material of the DNA) that occur in most cells in an animal's body, to half the normal number. Only after fertilization by sperm (also with the split half number of chromosomes) is the full complement restored. In parthenogenesis, no reduction splitting occurs, all eggs are full-chromosome clones of the mother, and all will also be female.

# Clearwing

The clearwing was so sparsely scaled
That panic was often entailed.
It looked like a bee
To all that could see.
Identification had failed.

Clearwing moths (family Sesiidae) are so called because they have clear, transparent wings, hardly marked by the coloured scales that give most butterflies and moths their pretty colours and patterns. This adaptation makes many of them look rather bee- or wasp-like, and the deception is emphasized by black and yellow banding on the body. Clearwings are day-flying moths, actively visiting flowers in bright sunshine – another very bee- or wasp-like trait.

# Assassin bug

Assassin sat quiet and still,
Waiting to make its first kill.
It stuck in its snout,
And sucked innards out,
Until it had eaten its fill.

Assassin bugs, family Reduviidae, are fierce predatory insects. They stalk their prey or sit and wait for hapless victims to come close enough to be pounced upon. Their nymphs also disguise themselves by sticking bits of rubbish, faeces and the remains of their dead prey onto themselves to break up their outline and merge into the background.

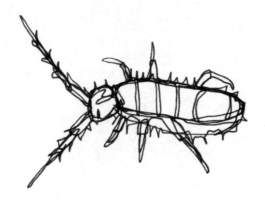

# Springtail

Having a spring in its tail
Helped the collembolan sail
Away through the air,
Spinning here, spinning there,
A whole-body uncontrolled flail.

Springtails, order Collembola, have a spring-loaded tail; the final segments, the furcula, like a minute tuning fork, are tucked underneath the abdomen where they are held in place by a clip organ called the retinaculum. If the animal is disturbed it tenses the spring, then releases the furcula which smacks onto the substrate – sending the collembolan spinning through the air, and hopefully out of danger.

Springtails, are now reckoned to be 'non-insect hexapods' based on obscure mouthpart characters barely visible even under very high-power microscopy. But until recently they have traditionally been included in the class Insecta, they always appear in insect monographs and textbooks, and they're insect enough for me.

# Appendix

This book started as an exercise in frivolity. It was only at about fifty or so limericks in that the formal idea of a book started to crystallize and the focus on insects became fixed. By that point, though, we had already touched on various other animal groups.

Those that follow are definitely not insects, and although technically 'entomologists' would not normally concern themselves with these organisms, most people interested in invertebrates are actually quite relaxed about which groups they study.

With that in mind, these last few entries have been rescued from the cutting room floor.

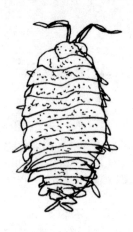

# Woodlouse

When counting the legs of woodlice,
It pays you to double check twice.
Though born with a dozen,
This crustacean cousin,
Has two more, fourteen, in a trice.

Woodlice (order Isopoda), the only truly terrestrial group of crustaceans, usually have fourteen legs, two on each of the seven central body sections. But when they are born, they have twelve. The female woodlouse carries her eggs in a pouch between her legs – the marsupium. They hatch here into tiny 12-legged larvae called mancas. Within a few hours the manca moults its skin, revealing an extra body segment, and thus two more legs, to give it the full complement of 14. The tiny woodlice are then released by the mother to live their independent lives.

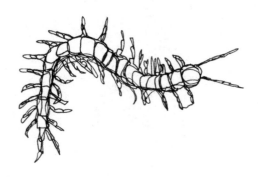

# Centipede

A centipede once bit a slug,
Which he foolishly thought was a bug.
His scimitar jaws
Were no good for gnaws,
As the slime made his mandibles yuck.

Centipedes (class Chilopoda) are multi-legged predatory arthropods with two legs on each body segment. The front pair of legs are modified into jaws and in many large species these can be seen as long curving fangs reaching around to the front of the head. These 'jaws' are hollow and contain venom sacks to kill their prey. Large species can pierce human skin to give a painful sting. Centipedes hunt all manner of small soil-dwelling invertebrates, although large slimy slugs are less attractive than small succulent insects.

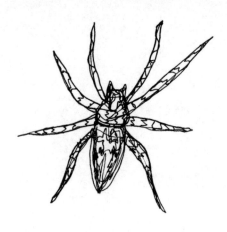

# Wolf spider

A wolf spider carried about,
Her young on her back, there's no doubt.
But when her strength left
And her babes were bereft,
They ate her until she was nowt.

Named for the habit of running about across the soil in large groups, wolf spiders (family Lycosidae) do not really hunt in packs but are simply gathering together in warm sunny places – often to bask in the sunshine or to find mates. The female carries her eggs about on her tail in a silk bag, and when they hatch she also carries the spiderlings on her body. There is evidence that they get protection from predators (especially other wolf spiders) and possibly take some nutrition from their mother's left-over prey items. But eventually the spiderlings leave to carry on their independent lives and are not above cannibalizing their dead parent.

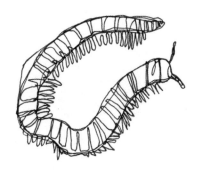

# Millipede

The millipede counted its toes,
As its legs were lined up in rows.
It got to five hundred,
But knew it had blundered,
By counting its arms and its nose.

Millipedes (class Diplopoda) are characterized by having four legs (two pairs) on each body segment, where centipedes have just two legs per segment. Despite common misconception, 100 legs do not make a centipede and no millipede ever has 1,000. In fact they broadly overlap in numbers, with some long centipedes reaching 400 legs and some short millipedes having just 36 limbs. Whereas centipedes are predators after small soil-dwelling invertebrates, millipedes are herbivores or detritivores. The maximum legs on any millipede is about 750.

# Index